Riddle 1

Why should you never tell secrets in a cornfield?

$\overline{4}\ \overline{8}\ \overline{17}\ \overline{14}\ \overline{10}\ \overline{11}\ \overline{8}$ $\overline{7}\ \overline{9}\ \overline{8}$ $\overline{17}\ \overline{16}\ \overline{12}\ \overline{13}$

$\overline{9}\ \overline{14}\ \overline{11}$ $\overline{8}\ \overline{14}\ \overline{12}\ \overline{11}$.

$\begin{array}{r}5\\+9\\\hline\end{array}$ =A $\begin{array}{r}0\\+4\\\hline\end{array}$ =B $\begin{array}{r}9\\+8\\\hline\end{array}$ =C $\begin{array}{r}6\\+2\\\hline\end{array}$ =E $\begin{array}{r}2\\+7\\\hline\end{array}$ =H

$\begin{array}{r}8\\+5\\\hline\end{array}$ =N $\begin{array}{r}10\\+6\\\hline\end{array}$ =O $\begin{array}{r}4\\+8\\\hline\end{array}$ =R $\begin{array}{r}1\\+10\\\hline\end{array}$ =S $\begin{array}{r}7\\+0\\\hline\end{array}$ =T

$\begin{array}{r}3\\+7\\\hline\end{array}$ =U

Addition and Subtraction Riddles Workbook

Riddle 2

Why is a baby like an old car?

$\overline{12}\ \overline{14}\ \overline{11}\ \overline{6}$ $\overline{10}\ \overline{5}\ \overline{12}\ \overline{14}$

$\overline{14}\ \overline{17}\ \overline{18}\ \overline{11}$ $\overline{17}$ $\overline{7}\ \overline{17}\ \overline{12}\ \overline{12}\ \overline{9}\ \overline{11}$.

$\begin{array}{r}10\\+7\\\hline\end{array}$ =A $\begin{array}{r}4\\+6\\\hline\end{array}$ =B $\begin{array}{r}2\\+9\\\hline\end{array}$ =E $\begin{array}{r}9\\+5\\\hline\end{array}$ =H $\begin{array}{r}6\\+3\\\hline\end{array}$ =L

$\begin{array}{r}1\\+4\\\hline\end{array}$ =O $\begin{array}{r}7\\+0\\\hline\end{array}$ =R $\begin{array}{r}3\\+9\\\hline\end{array}$ =T $\begin{array}{r}8\\+10\\\hline\end{array}$ =V $\begin{array}{r}5\\+1\\\hline\end{array}$ =Y

Addition and Subtraction Riddles Workbook

Riddle 3

What animal keeps the best time?

$\overline{10}$ $\overline{13}$ $\overline{10}$ $\overline{17}$ $\overline{19}$ $\overline{6}$ $\overline{12}$ $\overline{8}$ $\overline{11}$ $\overline{12}$ $\overline{8}$ $\overline{2}$ $\overline{9}$.

```
   2          9          3          1          6
  +8        +10         +9         +1         +5
  ――         ――         ――         ――         ――
  =A         =C         =D         =E         =G

   4          8          5         10          7
  +2         +0         +4         +7         +6
  ――         ――         ――         ――         ――
  =H         =O         =S         =T         =W
```

Addition and Subtraction Riddles Workbook

Riddle 4

What did one ghost say to the other ghost?

"$\overline{11}\ \overline{5}\quad \overline{6}\ \overline{5}\ \overline{0}\quad \overline{3}\ \overline{12}\ \overline{7}\ \overline{18}\ \overline{12}\ \overline{13}\ \overline{12}$

$\overline{18}\ \overline{17}\quad \overline{16}\ \overline{12}\ \overline{5}\ \overline{16}\ \overline{7}\ \overline{12}\,?$"

$\begin{array}{c}1\\+2\\\hline =B\end{array}\qquad \begin{array}{c}4\\+7\\\hline =D\end{array}\qquad \begin{array}{c}6\\+6\\\hline =E\end{array}\qquad \begin{array}{c}10\\+8\\\hline =I\end{array}\qquad \begin{array}{c}3\\+4\\\hline =L\end{array}\qquad \begin{array}{c}7\\+10\\\hline =N\end{array}$

$\begin{array}{c}2\\+3\\\hline =O\end{array}\qquad \begin{array}{c}9\\+7\\\hline =P\end{array}\qquad \begin{array}{c}0\\+0\\\hline =U\end{array}\qquad \begin{array}{c}8\\+5\\\hline =V\end{array}\qquad \begin{array}{c}5\\+1\\\hline =Y\end{array}$

Addition and Subtraction Riddles Workbook

Riddle 5

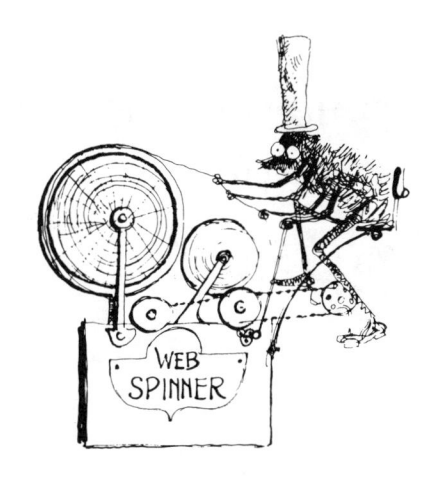

Why does a spider spin a web?

WEB SPINNER

$\overline{10}\ \overline{18}\ \overline{9}\ \overline{6}\ \overline{5}\ \overline{7}\ \overline{18}\ \ \overline{11}\ \overline{18}\ \ \overline{9}\ \overline{6}\ \overline{15}\ \overline{4}$,

$\overline{13}\ \overline{15}\ \overline{3}\ \overline{4}$.

$\begin{array}{r}3\\+3\\\hline\end{array}$ =A $\begin{array}{r}6\\+4\\\hline\end{array}$ =B $\begin{array}{r}9\\+0\\\hline\end{array}$ =C $\begin{array}{r}10\\+8\\\hline\end{array}$ =E $\begin{array}{r}7\\+4\\\hline\end{array}$ =H $\begin{array}{r}2\\+1\\\hline\end{array}$ =I

$\begin{array}{r}4\\+9\\\hline\end{array}$ =K $\begin{array}{r}8\\+7\\\hline\end{array}$ =N $\begin{array}{r}5\\+2\\\hline\end{array}$ =S $\begin{array}{r}1\\+3\\\hline\end{array}$ =T $\begin{array}{r}0\\+5\\\hline\end{array}$ =U

Addition and Subtraction Riddles Workbook

Riddle 6

A hippopotamus is sitting in a chair. What time is it?

$\overline{13}\ \overline{14}\ \overline{12}\ \overline{15}\ \ \overline{13}\ \overline{3}\ \ \overline{11}\ \overline{15}\ \overline{13}\ \ \overline{17}$

$\overline{10}\ \overline{15}\ \overline{9}\ \ \overline{6}\ \overline{8}\ \overline{17}\ \overline{14}\ \overline{16}$.

$\begin{array}{r}8\\+9\\\hline\end{array}$ =A $\begin{array}{r}0\\+6\\\hline\end{array}$ =C $\begin{array}{r}10\\+5\\\hline\end{array}$ =E $\begin{array}{r}3\\+8\\\hline\end{array}$ =G $\begin{array}{r}1\\+7\\\hline\end{array}$ =H

$\begin{array}{r}4\\+10\\\hline\end{array}$ =I $\begin{array}{r}9\\+3\\\hline\end{array}$ =M $\begin{array}{r}5\\+5\\\hline\end{array}$ =N $\begin{array}{r}2\\+1\\\hline\end{array}$ =O $\begin{array}{r}6\\+10\\\hline\end{array}$ =R

$\begin{array}{r}7\\+6\\\hline\end{array}$ =T $\begin{array}{r}0\\+9\\\hline\end{array}$ =W

Addition and Subtraction Riddles Workbook

Riddle 7

Why are soldiers and dentists alike?

$\overline{9}\ \overline{17}\ \overline{12}\ \overline{5}$ $\overline{13}\ \overline{19}\ \overline{9}\ \overline{17}$ $\overline{17}\ \overline{8}\ \overline{11}\ \overline{12}$

$\overline{9}\ \overline{19}$ $\overline{15}\ \overline{7}\ \overline{14}\ \overline{10}\ \overline{10}$.

$\begin{array}{c} 2 \\ +6 \\ \hline =A \end{array}$ $\begin{array}{c} 5 \\ +8 \\ \hline =B \end{array}$ $\begin{array}{c} 9 \\ +6 \\ \hline =D \end{array}$ $\begin{array}{c} 3 \\ +9 \\ \hline =E \end{array}$ $\begin{array}{c} 7 \\ +10 \\ \hline =H \end{array}$ $\begin{array}{c} 8 \\ +6 \\ \hline =I \end{array}$

$\begin{array}{c} 0 \\ +10 \\ \hline =L \end{array}$ $\begin{array}{c} 10 \\ +9 \\ \hline =O \end{array}$ $\begin{array}{c} 6 \\ +1 \\ \hline =R \end{array}$ $\begin{array}{c} 1 \\ +8 \\ \hline =T \end{array}$ $\begin{array}{c} 4 \\ +7 \\ \hline =V \end{array}$ $\begin{array}{c} 2 \\ +3 \\ \hline =Y \end{array}$

Addition and Subtraction Riddles Workbook

Riddle 8

What kind of cat crawls like a snake?

$\overline{6}$ $\overline{16}$ $\overline{12}$ $\overline{18}$ $\overline{15}$ $\overline{13}$ $\overline{11}$ $\overline{9}$ $\overline{5}$ $\overline{3}$ $\overline{10}$ $\overline{8}$.

$\begin{array}{r}2\\+4\\\hline\end{array}$=A $\begin{array}{r}5\\+7\\\hline\end{array}$=A $\begin{array}{r}1\\+9\\\hline\end{array}$=A $\begin{array}{r}10\\+6\\\hline\end{array}$=C $\begin{array}{r}7\\+8\\\hline\end{array}$=E $\begin{array}{r}7\\+2\\\hline\end{array}$=I

$\begin{array}{r}4\\+1\\\hline\end{array}$=L $\begin{array}{r}0\\+3\\\hline\end{array}$=L $\begin{array}{r}6\\+5\\\hline\end{array}$=P $\begin{array}{r}8\\+0\\\hline\end{array}$=R $\begin{array}{r}3\\+10\\\hline\end{array}$=R $\begin{array}{r}9\\+9\\\hline\end{array}$=T

Addition and Subtraction Riddles Workbook

Riddle 9

What kinds of keys won't open a door?

$\overline{12}\ \overline{9}\ \overline{3}\ \overline{8}\ \overline{4}\ \overline{14}\ \overline{16}$, $\overline{17}\ \overline{6}\ \overline{19}\ \overline{8}\ \overline{4}\ \overline{14}\ \overline{16}$,

$\overline{18}\ \overline{3}\ \overline{13}$ $\overline{13}\ \overline{9}\ \overline{3}\ \overline{8}\ \overline{4}\ \overline{14}\ \overline{16}$.

$\begin{array}{c}9\\+9\\\hline\end{array}$=A $\begin{array}{c}6\\+7\\\hline\end{array}$=D $\begin{array}{c}4\\+0\\\hline\end{array}$=E $\begin{array}{c}3\\+5\\\hline\end{array}$=K $\begin{array}{c}2\\+10\\\hline\end{array}$=M $\begin{array}{c}0\\+3\\\hline\end{array}$=N

$\begin{array}{c}5\\+4\\\hline\end{array}$=O $\begin{array}{c}9\\+10\\\hline\end{array}$=R $\begin{array}{c}7\\+9\\\hline\end{array}$=S $\begin{array}{c}10\\+7\\\hline\end{array}$=T $\begin{array}{c}4\\+2\\\hline\end{array}$=U $\begin{array}{c}8\\+6\\\hline\end{array}$=Y

Addition and Subtraction Riddles Workbook

Riddle 10

What coat is put on only when it is wet?

$\overline{7}$ $\overline{11}\,\overline{14}\,\overline{12}\,\overline{16}$ $\overline{10}\,\overline{15}$ $\overline{3}\,\overline{6}\,\overline{8}\,\overline{13}\,\overline{9}$.

$\begin{array}{r}7\\+5\\\hline\end{array}$ =A

$\begin{array}{r}3\\+4\\\hline\end{array}$ =A

$\begin{array}{r}5\\+1\\\hline\end{array}$ =A

$\begin{array}{r}9\\+2\\\hline\end{array}$ =C

$\begin{array}{r}6\\+9\\\hline\end{array}$ =F

$\begin{array}{r}1\\+7\\\hline\end{array}$ =I

$\begin{array}{r}10\\+3\\\hline\end{array}$ =N

$\begin{array}{r}2\\+8\\\hline\end{array}$ =O

$\begin{array}{r}4\\+10\\\hline\end{array}$ =O

$\begin{array}{r}3\\+0\\\hline\end{array}$ =P

$\begin{array}{r}8\\+8\\\hline\end{array}$ =T

$\begin{array}{r}5\\+4\\\hline\end{array}$ =T

Addition and Subtraction Riddles Workbook

Riddle 11

What letters are not in the alphabet?

$\overline{16}\ \overline{6}\ \overline{11}$ $\overline{17}\ \overline{10}\ \overline{11}\ \overline{9}$ $\overline{12}\ \overline{10}$ $\overline{16}\ \overline{6}\ \overline{11}$

$\overline{7}\ \overline{15}\ \overline{12}\ \overline{20}\ \overline{1}\ \overline{17}\ \overline{2}$.

$\begin{array}{r}7\\+8\\\hline\end{array}$=A $\begin{array}{r}1\\+0\\\hline\end{array}$=B $\begin{array}{r}6\\+5\\\hline\end{array}$=E $\begin{array}{r}2\\+4\\\hline\end{array}$=H $\begin{array}{r}3\\+9\\\hline\end{array}$=I $\begin{array}{r}10\\+10\\\hline\end{array}$=L

$\begin{array}{r}4\\+3\\\hline\end{array}$=M $\begin{array}{r}5\\+5\\\hline\end{array}$=N $\begin{array}{r}9\\+8\\\hline\end{array}$=O $\begin{array}{r}8\\+1\\\hline\end{array}$=S $\begin{array}{r}10\\+6\\\hline\end{array}$=T $\begin{array}{r}0\\+2\\\hline\end{array}$=X

Addition and Subtraction Riddles Workbook

Riddle 12

When can you carry water in a strainer?

$\overline{15}\ \overline{11}\ \overline{20}\ \overline{2}\quad \overline{16}\ \overline{1}\ \overline{8}\ ,\quad \overline{9}\ \overline{7}\ \overline{12}\ \overline{5}\ \overline{10}\ \overline{14}$.

$\begin{array}{r}10\\+10\\\hline\end{array}$ =E

$\begin{array}{r}5\\+5\\\hline\end{array}$ =E

$\begin{array}{r}7\\+2\\\hline\end{array}$ =F

$\begin{array}{r}3\\+8\\\hline\end{array}$ =H

$\begin{array}{r}6\\+10\\\hline\end{array}$ =I

$\begin{array}{r}9\\+5\\\hline\end{array}$ =N

$\begin{array}{r}2\\+0\\\hline\end{array}$ =N

$\begin{array}{r}8\\+4\\\hline\end{array}$ =O

$\begin{array}{r}1\\+6\\\hline\end{array}$ =R

$\begin{array}{r}4\\+4\\\hline\end{array}$ =S

$\begin{array}{r}0\\+1\\\hline\end{array}$ =T

$\begin{array}{r}7\\+8\\\hline\end{array}$ =W

$\begin{array}{r}3\\+2\\\hline\end{array}$ =Z

Addition and Subtraction Riddles Workbook

Riddle 13

What colors would you paint the sun and wind?

$\overline{6}\ \overline{4}\ \overline{16}$ $\overline{7}\ \overline{12}\ \overline{13}$ $\overline{18}\ \overline{2}\ \overline{7}\ \overline{16}$ $\overline{10}\ \overline{13}\ \overline{8}$

$\overline{6}\ \overline{4}\ \overline{16}$ $\overline{14}\ \overline{19}\ \overline{13}\ \overline{8}$ $\overline{9}\ \overline{11}\ \overline{12}\ \overline{16}$.

$\begin{array}{c}4\\+6\\\hline\end{array}$ =A $\begin{array}{c}7\\+2\\\hline\end{array}$ =B $\begin{array}{c}5\\+3\\\hline\end{array}$ =D $\begin{array}{c}8\\+8\\\hline\end{array}$ =E $\begin{array}{c}3\\+1\\\hline\end{array}$ =H $\begin{array}{c}9\\+10\\\hline\end{array}$ =I

$\begin{array}{c}2\\+9\\\hline\end{array}$ =L $\begin{array}{c}6\\+7\\\hline\end{array}$ =N $\begin{array}{c}1\\+1\\\hline\end{array}$ =O $\begin{array}{c}10\\+8\\\hline\end{array}$ =R $\begin{array}{c}7\\+0\\\hline\end{array}$ =S $\begin{array}{c}4\\+2\\\hline\end{array}$ =T

$\begin{array}{c}8\\+4\\\hline\end{array}$ =U $\begin{array}{c}5\\+9\\\hline\end{array}$ =W

Addition and Subtraction Riddles Workbook

Riddle 14

What did the tall chimney say to the short chimney?

" __ __ __ __ __   __ __ __   __ __ __ ,
  6 17 8 10 14   11 17 2   9 13 3

  __ __ __ __ __ __   __ __   __ __ __ __ __ ."
  14 11 17 8 3 12   2 17   15 20 17 7 14

$\begin{array}{r}3\\+6\\\hline\end{array}$ =B $\begin{array}{r}5\\+9\\\hline\end{array}$ =E $\begin{array}{r}1\\+2\\\hline\end{array}$ =G $\begin{array}{r}4\\+8\\\hline\end{array}$ =H $\begin{array}{r}6\\+7\\\hline\end{array}$ =I

$\begin{array}{r}2\\+5\\\hline\end{array}$ =K $\begin{array}{r}10\\+10\\\hline\end{array}$ =M $\begin{array}{r}7\\+4\\\hline\end{array}$ =N $\begin{array}{r}9\\+8\\\hline\end{array}$ =O $\begin{array}{r}0\\+10\\\hline\end{array}$ =R

$\begin{array}{r}8\\+7\\\hline\end{array}$ =S $\begin{array}{r}1\\+1\\\hline\end{array}$ =T $\begin{array}{r}6\\+2\\\hline\end{array}$ =U $\begin{array}{r}3\\+3\\\hline\end{array}$ =Y

Addition and Subtraction Riddles Workbook

Riddle 15

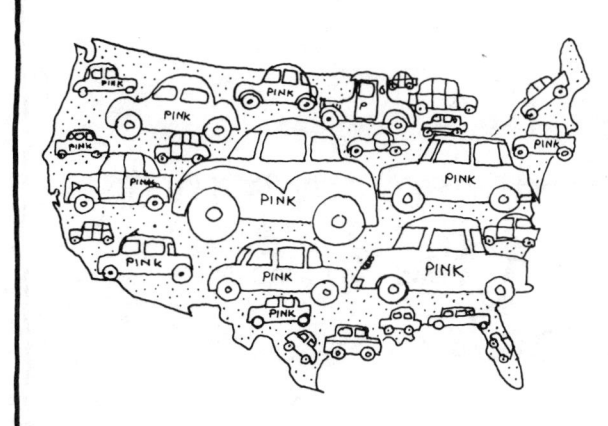

If everyone in the nation owned pink cars, what would the nation be called?

$\overline{9}$ $\overline{10}\,\overline{11}\,\overline{8}\,\overline{5}$ $\overline{12}\,\overline{15}\,\overline{7}\,\overline{16}\,\overline{6}\,\overline{18}\,\overline{14}\,\overline{13}\,\overline{3}$.

$\begin{array}{c}10\\+5\\\hline=A\end{array}$ $\begin{array}{c}6\\+3\\\hline=A\end{array}$ $\begin{array}{c}2\\+4\\\hline=A\end{array}$ $\begin{array}{c}9\\+3\\\hline=C\end{array}$ $\begin{array}{c}5\\+6\\\hline=I\end{array}$ $\begin{array}{c}7\\+7\\\hline=I\end{array}$

$\begin{array}{c}4\\+1\\\hline=K\end{array}$ $\begin{array}{c}8\\+0\\\hline=N\end{array}$ $\begin{array}{c}9\\+7\\\hline=N\end{array}$ $\begin{array}{c}1\\+2\\\hline=N\end{array}$ $\begin{array}{c}3\\+10\\\hline=O\end{array}$ $\begin{array}{c}9\\+1\\\hline=P\end{array}$

$\begin{array}{c}0\\+7\\\hline=R\end{array}$ $\begin{array}{c}10\\+8\\\hline=T\end{array}$

Addition and Subtraction Riddles Workbook

Riddle 16

What are cows that sit on the grass called?

$\overline{7}\ \overline{4}\ \overline{1}\ \overline{3}\ \overline{5}\ \overline{10}\ \ \overline{0}\ \overline{6}\ \overline{8}\ \overline{2}.$

$\begin{array}{r}7\\-7\\\hline\end{array}$ =B $\begin{array}{r}13\\-3\\\hline\end{array}$ =D $\begin{array}{r}17\\-9\\\hline\end{array}$ =E $\begin{array}{r}14\\-8\\\hline\end{array}$ =E $\begin{array}{r}3\\-1\\\hline\end{array}$ =F

$\begin{array}{r}9\\-2\\\hline\end{array}$ =G $\begin{array}{r}11\\-6\\\hline\end{array}$ =N $\begin{array}{r}5\\-4\\\hline\end{array}$ =O $\begin{array}{r}4\\-0\\\hline\end{array}$ =R $\begin{array}{r}8\\-5\\\hline\end{array}$ =U

Addition and Subtraction Riddles Workbook

Riddle 17

How do locomotives hear?

$\overline{6}\ \overline{5}\ \overline{8}\ \overline{10}\ \overline{4}\ \overline{3}\ \overline{5}\quad \overline{6}\ \overline{5}\ \overline{9}\ \overline{1}\ \overline{8}$

$\overline{9}\ \overline{2}\ \overline{3}\ \overline{1}\ \overline{2}\ \overline{9}\ \overline{9}\ \overline{8}\ \overline{7}$.

$\begin{array}{c}18\\-9\\\hline\end{array}$ =E $\begin{array}{c}7\\-4\\\hline\end{array}$ =G $\begin{array}{c}10\\-5\\\hline\end{array}$ =H $\begin{array}{c}8\\-7\\\hline\end{array}$ =I $\begin{array}{c}5\\-3\\\hline\end{array}$ =N

$\begin{array}{c}12\\-2\\\hline\end{array}$ =O $\begin{array}{c}9\\-1\\\hline\end{array}$ =R $\begin{array}{c}15\\-8\\\hline\end{array}$ =S $\begin{array}{c}12\\-6\\\hline\end{array}$ =T $\begin{array}{c}11\\-7\\\hline\end{array}$ =U

Addition and Subtraction Riddles Workbook

Riddle 18

What did the flower say to the bee?

"$\overline{\underset{9}{S}\,\underset{1}{T}\,\underset{4}{O}\,\underset{10}{P}}$ $\overline{\underset{2}{B}\,\underset{5}{U}\,\underset{0}{G}\,\underset{0}{G}\,\underset{7}{I}\,\underset{8}{N}\,\underset{0}{G}}$ $\overline{\underset{3}{M}\,\underset{6}{E}}$!"

8 − 6 = B	9 − 3 = E	8 − 8 = G	11 − 4 = I	13 − 10 = M	10 − 2 = N
13 − 9 = O	10 − 0 = P	16 − 7 = S	2 − 1 = T	10 − 5 = U	

Addition and Subtraction Riddles Workbook

Riddle 19

What did one candle say to the other candle?

"$\overline{9\ 8\ 0}$ $\overline{5\ 1\ 3}$ $\overline{11\ 1\ 6\ 2\ 11}$ $\overline{1\ 3\ 7}$ $\overline{7\ 1\ 2\ 6\ 11\ 4\ 7}$?"

10 4 11 10 16 9
-1 -4 -0 -6 -10 -7
=A =E =G =H =I =N

3 16 12 6 14
-2 -8 -5 -3 -9
=O =R =T =U =Y

Addition and Subtraction Riddles Workbook

Riddle 20

Where can everyone always find money when they look for it?

$\overline{4}\ \overline{3}\quad \overline{7}\ \overline{10}\ \overline{5}\quad \overline{1}\ \overline{4}\ \overline{6}\ \overline{7}\ \overline{4}\ \overline{13}\ \overline{3}\ \overline{8}\ \overline{2}\ \overline{9}$.

$\begin{array}{r}13\\-5\\\hline\end{array}$ =A $\begin{array}{r}7\\-1\\\hline\end{array}$ =C $\begin{array}{r}4\\-3\\\hline\end{array}$ =D $\begin{array}{r}15\\-10\\\hline\end{array}$ =E $\begin{array}{r}14\\-4\\\hline\end{array}$ =H $\begin{array}{r}12\\-8\\\hline\end{array}$ =I

$\begin{array}{r}10\\-7\\\hline\end{array}$ =N $\begin{array}{r}15\\-2\\\hline\end{array}$ =O $\begin{array}{r}11\\-9\\\hline\end{array}$ =R $\begin{array}{r}13\\-6\\\hline\end{array}$ =T $\begin{array}{r}16\\-7\\\hline\end{array}$ =Y

Addition and Subtraction Riddles Workbook

Riddle 21

What did one wall say to the other wall?

"LET'S MEET AT THE CORNER."

Problem	= Letter
7 − 3 = 4	A
16 − 6 = 10	C
9 − 4 = 5	E
8 − 7 = 1	H
14 − 5 = 9	L
5 − 2 = 3	M
8 − 1 = 7	N
10 − 8 = 2	O
6 − 0 = 6	R
18 − 10 = 8	S
9 − 9 = 0	T

Riddle 22

What time is it when the clock strikes thirteen?

$\overline{6}\;\overline{4}\;\overline{9}\;\overline{10}$ $\overline{6}\;\overline{13}$ $\overline{3}\;\overline{4}\;\overline{1}$ $\overline{6}\;\overline{2}\;\overline{10}$

$\overline{0}\;\overline{5}\;\overline{13}\;\overline{0}\;\overline{11}$.

$\begin{array}{r}3\\-3\\\hline\end{array}$ =C
$\begin{array}{r}15\\-5\\\hline\end{array}$ =E
$\begin{array}{r}11\\-8\\\hline\end{array}$ =F
$\begin{array}{r}4\\-2\\\hline\end{array}$ =H
$\begin{array}{r}11\\-7\\\hline\end{array}$ =I
$\begin{array}{r}12\\-1\\\hline\end{array}$ =K

$\begin{array}{r}16\\-11\\\hline\end{array}$ =L
$\begin{array}{r}18\\-9\\\hline\end{array}$ =M
$\begin{array}{r}17\\-4\\\hline\end{array}$ =O
$\begin{array}{r}12\\-6\\\hline\end{array}$ =T
$\begin{array}{r}1\\-0\\\hline\end{array}$ =X

Addition and Subtraction Riddles Workbook

Riddle 23

What's the difference between a watchmaker and a prison warden?

__ __ __   __ __ __ __ __   __ __ __ __ __ __ __ ,
11 7 4   12 4 1 1 12   8 2 3 9 10 4 12

__ __ __   __ __ __   __ __ __ __ __
2 7 6   3 10 4   11 3 10 4 5

__ __ __ __ __ __ __   __ __ __ __ __ .
8 2 3 9 10 4 12   9 4 1 1 12

| 10
−8
=A | 11
−2
=C | 18
−12
=D | 14
−10
=E | 13
−3
=H | 7
−6
=L |

| 7
−0
=N | 15
−4
=O | 16
−11
=R | 17
−5
=S | 12
−9
=T | 9
−1
=W |

Addition and Subtraction Riddles Workbook

Riddle 24

What has 18 legs and catches flies?

$\overline{3}$ $\overline{0}\,\overline{5}\,\overline{9}\,\overline{8}\,\overline{11}\,\overline{1}\,\overline{7}\,\overline{4}$ $\overline{6}\,\overline{2}\,\overline{10}\,\overline{12}$

$\begin{array}{r}12\\-7\\\hline\end{array}$ =A $\begin{array}{r}10\\-0\\\hline\end{array}$ =A $\begin{array}{r}5\\-4\\\hline\end{array}$ =A $\begin{array}{r}9\\-6\\\hline\end{array}$ =A $\begin{array}{r}12\\-1\\\hline\end{array}$ =B $\begin{array}{r}3\\-3\\\hline\end{array}$ =B

$\begin{array}{r}7\\-5\\\hline\end{array}$ =E $\begin{array}{r}18\\-10\\\hline\end{array}$ =E $\begin{array}{r}11\\-7\\\hline\end{array}$ =L $\begin{array}{r}16\\-9\\\hline\end{array}$ =L $\begin{array}{r}14\\-2\\\hline\end{array}$ =M $\begin{array}{r}20\\-11\\\hline\end{array}$ =S

$\begin{array}{r}14\\-8\\\hline\end{array}$ =T

Addition and Subtraction Riddles Workbook

Riddle 25

Why do dragons sleep in the daytime?

$\overline{2}\ \overline{9}$ $\overline{6}\ \overline{4}\ \overline{14}\ \overline{11}$ $\overline{3}\ \overline{0}\ \overline{7}$ $\overline{4}\ \overline{5}\ \overline{7}\ \overline{6}$

$\overline{10}\ \overline{7}\ \overline{1}\ \overline{8}\ \overline{4}\ \overline{6}\ \overline{2}$.

$\dfrac{6}{-6}$ = A $\dfrac{12}{-9}$ = C $\dfrac{21}{-7}$ = E $\dfrac{17}{-9}$ = G $\dfrac{9}{-5}$ = H

$\dfrac{3}{-2}$ = I $\dfrac{11}{-1}$ = K $\dfrac{14}{-7}$ = N $\dfrac{15}{-6}$ = O $\dfrac{7}{-5}$ = S

$\dfrac{10}{-4}$ = T $\dfrac{8}{-3}$ = U $\dfrac{44}{-33}$ = Y

Addition and Subtraction Riddles Workbook

Riddle 26

What did the mayonnaise say to the refrigerator?

"CLOSE THE DOOR, I'M DRESSING."

$15 - 8 = 7 = C$

$9 - 6 = 3 = D$

$19 - 10 = 9 = E$

$14 - 3 = 11 = G$

$8 - 7 = 1 = H$

$5 - 1 = 4 = I$

$21 - 11 = 10 = L$

$17 - 4 = 13 = M$

$4 - 2 = 2 = N$

$17 - 9 = 8 = O$

$6 - 0 = 6 = R$

$24 - 12 = 12 = S$

$5 - 5 = 0 = T$

Riddle 27

What's the hardest thing about learning to ride a bicycle?

$\overline{1\ 9\ 4}$ $\overline{11\ 8\ 0\ 4\ 5\ 4\ 7\ 1}$ $\overline{2\ 10}$

$\overline{1\ 9\ 4}$ $\overline{9\ 8\ 6\ 3\ 4\ 10\ 1}$ $\overline{1\ 9\ 2\ 7\ 13}$

$\underline{18}\\-10$ =A

$\underline{9}\\-6$ =D

$\underline{11}\\-7$ =E

$\underline{15}\\-2$ =G

$\underline{16}\\-7$ =H

$\underline{10}\\-8$ =I

$\underline{13}\\-8$ =M

$\underline{15}\\-8$ =N

$\underline{12}\\-1$ =P

$\underline{15}\\-9$ =R

$\underline{13}\\-3$ =S

$\underline{5}\\-4$ =T

$\underline{5}\\-5$ =V

Addition and Subtraction Riddles Workbook

Riddle 28

What did one hair say to the other?

"$\overline{7}$ $\overline{4}\,\overline{11}\,\overline{5}\,\overline{1}$, $\overline{1}\,\overline{6}\,\overline{2}\,\overline{2}$ $\overline{0}\,\overline{5}\,\overline{5}\,\overline{11}$ $\overline{8}\,\overline{6}$ $\overline{10}\,\overline{3}\,\overline{13}\,\overline{9}\,\overline{6}\,\overline{12}$."

(I know, we'll soon be parted.)

7 − 4 = A	9 − 1 = B	20 − 8 = D	13 − 7 = E	17 − 10 = I	17 − 13 = K
13 − 11 = L	11 − 0 = N	7 − 2 = O	22 − 12 = P	22 − 9 = R	5 − 5 = S
12 − 3 = T	7 − 6 = W				

Addition and Subtraction Riddles Workbook

1.

Why should you never tell secrets in a cornfield?

BECAUSE THE CORN HAS EARS.

4 8 17 14 10 11 8 / 7 9 8 / 17 16 12 13 / 9 14 11 / 8 14 12 11

$5+9=14$ = A; $0+4=4$ = B; $9+8=17$ = C; $6+2=8$ = E; $2+7=9$ = H; $8+5=13$ = N; $10+6=16$ = O; $4+8=12$ = R; $1+10=11$ = S; $7+0=7$ = T; $3+7=10$ = U

2.

Why is a baby like an old car?

THEY BOTH HAVE A RATTLE.

12 14 11 6 / 10 5 12 14 / 14 17 18 11 / 17 / 7 17 12 12 9 11

$10+7=17$ = A; $4+6=10$ = B; $2+9=11$ = E; $9+5=14$ = H; $6+3=9$ = L; $1+4=5$ = O; $7+0=7$ = R; $3+9=12$ = T; $8+10=18$ = V; $5+1=6$ = Y

3.

What animal keeps the best time?

A WATCHDOG DOES.

10 / 13 10 17 19 6 12 8 11 / 12 8 2 9

$2+8=10$ = A; $9+10=19$ = C; $3+9=12$ = D; $1+1=2$ = E; $6+5=11$ = G; $4+2=6$ = H; $8+0=8$ = O; $5+4=9$ = S; $10+7=17$ = T; $7+6=13$ = W

4.

What did one ghost say to the other ghost?

"DO YOU BELIEVE IN PEOPLE?"

11 5 / 6 5 0 / 3 12 7 18 12 13 12 / 18 17 / 16 12 5 16 7 12

$1+2=3$ = B; $4+7=11$ = D; $6+6=12$ = E; $10+8=18$ = I; $3+4=7$ = L; $7+10=17$ = N; $2+3=5$ = O; $9+7=16$ = P; $0+0=0$ = U; $8+5=3$ = V; $5+1=6$ = Y

5.

Why does a spider spin a web?

BECAUSE IT CAN'T KNIT.

10 18 9 6 5 7 18 / 11 18 / 9 6 15 4 / 13 15 3 4

$3+3=6$ = A; $6+4=10$ = B; $9+0=9$ = C; $10+8=18$ = E; $7+4=11$ = H; $2+1=3$ = I; $4+9=13$ = K; $8+7=15$ = N; $5+2=7$ = S; $1+3=4$ = T; $0+5=5$ = U

6.

A hippopotamus is sitting in a chair. What time is it?

TIME TO GET A NEW CHAIR.

13 14 12 15 / 13 3 / 11 15 13 / 17 / 10 15 9 / 6 8 17 14 16

$8+9=17$ = A; $0+6=6$ = C; $10+5=15$ = E; $3+8=11$ = G; $1+7=8$ = H; $4+10=14$ = I; $9+3=12$ = M; $5+5=10$ = N; $2+1=3$ = O; $6+10=16$ = R; $7+6=13$ = T; $0+9=9$ = W

7.

Why are soldiers and dentists alike?

THEY BOTH HAVE TO DRILL.

9 17 12 5 / 13 19 9 17 / 17 8 11 12 / 9 19 / 15 7 14 10 10

$2+6=8$ = A; $5+8=13$ = B; $9+6=15$ = D; $3+9=12$ = E; $7+10=17$ = H; $8+6=14$ = I; $0+10=10$ = L; $10+9=19$ = O; $6+1=7$ = R; $1+8=9$ = T; $4+7=11$ = V; $2+3=5$ = Y

8.

What kind of cat crawls like a snake?

A CATERPILLAR.

6 / 16 12 18 15 13 11 9 5 3 10 8

$2+4=6$ = A; $5+7=12$ = A; $1+9=10$ = A; $10+6=16$ = C; $7+8=15$ = E; $7+2=9$ = I; $4+1=5$ = L; $0+3=3$ = L; $6+5=11$ = P; $8+0=8$ = R; $3+10=13$ = R; $9+9=18$ = T

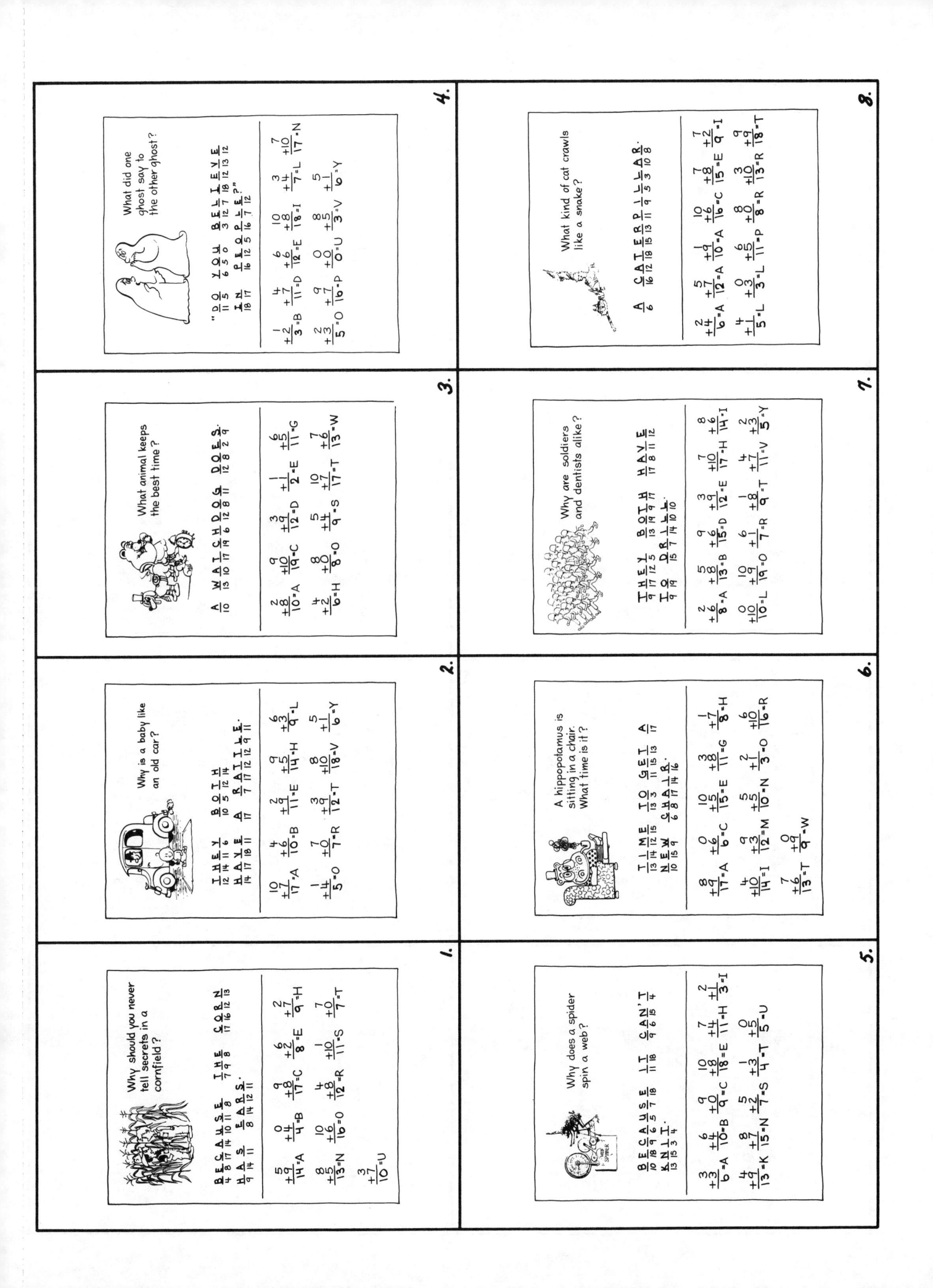

12.

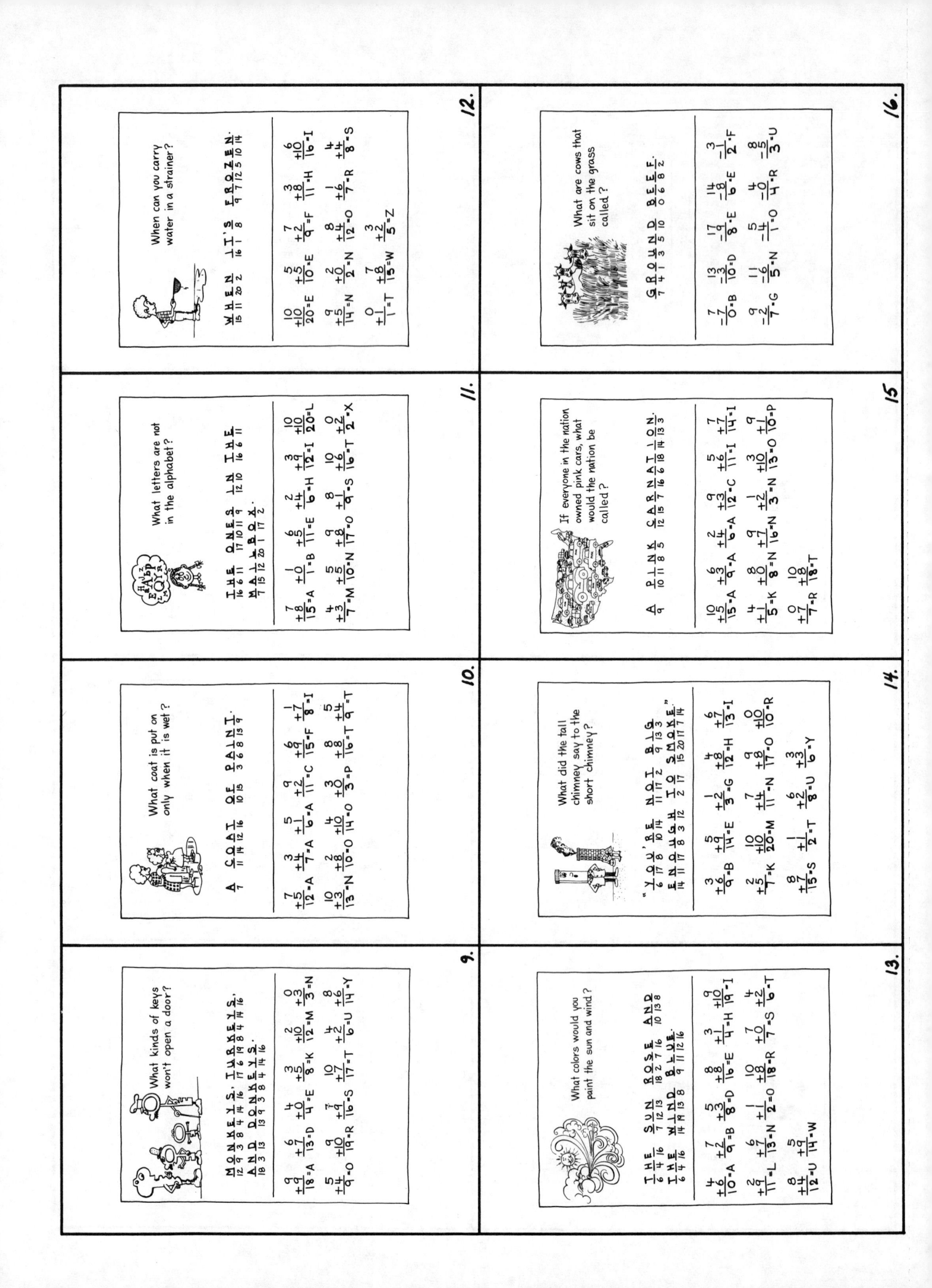

When can you carry water in a strainer?

W H E N   I T ' S   F R O Z E N
15 11 20 2   8   9 7 12 5 10 14

$\begin{array}{r}10\\+10\\\hline 20\end{array}$=E $\begin{array}{r}5\\+5\\\hline 10\end{array}$=E $\begin{array}{r}7\\+2\\\hline 9\end{array}$=F $\begin{array}{r}3\\+8\\\hline 11\end{array}$=H $\begin{array}{r}6\\+10\\\hline 16\end{array}$=I

$\begin{array}{r}9\\+5\\\hline 14\end{array}$=N $\begin{array}{r}2\\+0\\\hline 2\end{array}$=N $\begin{array}{r}8\\+12\\\hline 20\end{array}$=O $\begin{array}{r}6\\+1\\\hline 7\end{array}$=R $\begin{array}{r}4\\+8\\\hline 8\end{array}$=S

$\begin{array}{r}0\\+1\\\hline 1\end{array}$=T $\begin{array}{r}8\\+7\\\hline 15\end{array}$=W $\begin{array}{r}3\\+2\\\hline 5\end{array}$=Z

16.

What are cows that sit on the grass called?

G R O U N D   B E E F
7 4 1 3 5 10   0 6 8 2

$\begin{array}{r}7\\-7\\\hline 0\end{array}$=B $\begin{array}{r}13\\-3\\\hline 10\end{array}$=D $\begin{array}{r}17\\-9\\\hline 8\end{array}$=E $\begin{array}{r}14\\-8\\\hline 6\end{array}$=E $\begin{array}{r}3\\-1\\\hline 2\end{array}$=F

$\begin{array}{r}9\\-2\\\hline 7\end{array}$=G $\begin{array}{r}11\\-6\\\hline 5\end{array}$=N $\begin{array}{r}5\\-1\\\hline 4\end{array}$=N $\begin{array}{r}4\\-0\\\hline 4\end{array}$=R $\begin{array}{r}8\\-5\\\hline 3\end{array}$=U

11.

What letters are not in the alphabet?

T H E   O N E S   I N   T H E   M A I L B O X
16 6 11 17 10 11 9 12 10 16 6 11 7 15 12 20 1 17 2

$\begin{array}{r}7\\+8\\\hline 15\end{array}$=A $\begin{array}{r}1\\+0\\\hline 1\end{array}$=B $\begin{array}{r}6\\+5\\\hline 11\end{array}$=E $\begin{array}{r}2\\+7\\\hline 9\end{array}$=H $\begin{array}{r}3\\+9\\\hline 12\end{array}$=I $\begin{array}{r}10\\+10\\\hline 20\end{array}$=L

$\begin{array}{r}4\\+3\\\hline 7\end{array}$=M $\begin{array}{r}5\\+5\\\hline 10\end{array}$=N $\begin{array}{r}8\\+9\\\hline 17\end{array}$=O $\begin{array}{r}8\\+8\\\hline 16\end{array}$=S $\begin{array}{r}0\\+6\\\hline 6\end{array}$=T $\begin{array}{r}0\\+2\\\hline 2\end{array}$=X

15.

If everyone in the nation owned pink cars, what would the nation be called?

A   P I N K   C A R N A T I O N
9   10 11 8 5 12 15 7 16 6 18 14 13 3

$\begin{array}{r}10\\+5\\\hline 15\end{array}$=A $\begin{array}{r}6\\+3\\\hline 9\end{array}$=A $\begin{array}{r}2\\+4\\\hline 6\end{array}$=A $\begin{array}{r}9\\+3\\\hline 12\end{array}$=C $\begin{array}{r}5\\+6\\\hline 11\end{array}$=I $\begin{array}{r}7\\+7\\\hline 14\end{array}$=I

$\begin{array}{r}4\\+1\\\hline 5\end{array}$=K $\begin{array}{r}8\\+8\\\hline 16\end{array}$=N $\begin{array}{r}4\\+7\\\hline 9\end{array}$=N $\begin{array}{r}1\\+2\\\hline 3\end{array}$=N $\begin{array}{r}5\\+8\\\hline 13\end{array}$=O $\begin{array}{r}3\\+7\\\hline 10\end{array}$=P

$\begin{array}{r}0\\+7\\\hline 7\end{array}$=R $\begin{array}{r}10\\+8\\\hline 18\end{array}$=T

10.

What coat is put on only when it is wet?

A   C O A T   O F   P A I N T
7   11 14 12 16   10 15   3 6 8 13 9

$\begin{array}{r}7\\+5\\\hline 12\end{array}$=A $\begin{array}{r}3\\+4\\\hline 7\end{array}$=A $\begin{array}{r}5\\+2\\\hline 6\end{array}$=A $\begin{array}{r}9\\+2\\\hline 11\end{array}$=C $\begin{array}{r}6\\+9\\\hline 15\end{array}$=F $\begin{array}{r}7\\+8\\\hline 16\end{array}$=I

$\begin{array}{r}10\\+3\\\hline 13\end{array}$=N $\begin{array}{r}2\\+8\\\hline 10\end{array}$=N $\begin{array}{r}4\\+10\\\hline 14\end{array}$=O $\begin{array}{r}8\\+0\\\hline 8\end{array}$=O $\begin{array}{r}6\\+3\\\hline 9\end{array}$=P $\begin{array}{r}5\\+4\\\hline 9\end{array}$=T

14.

What did the tall chimney say to the short chimney?

" Y O U ' R E   N O T   B I G   E N O U G H   T O   S M O K E "
6 7 8 9 13 3   10 14 1   17 2   15 20 17 7 14

$\begin{array}{r}3\\+6\\\hline 9\end{array}$=B $\begin{array}{r}5\\+9\\\hline 14\end{array}$=E $\begin{array}{r}1\\+2\\\hline 3\end{array}$=G $\begin{array}{r}4\\+8\\\hline 12\end{array}$=H $\begin{array}{r}6\\+7\\\hline 13\end{array}$=I

$\begin{array}{r}2\\+5\\\hline 7\end{array}$=K $\begin{array}{r}10\\+10\\\hline 20\end{array}$=M $\begin{array}{r}7\\+1\\\hline 8\end{array}$=N $\begin{array}{r}9\\+8\\\hline 17\end{array}$=O $\begin{array}{r}0\\+10\\\hline 10\end{array}$=R

$\begin{array}{r}8\\+7\\\hline 15\end{array}$=S $\begin{array}{r}1\\+1\\\hline 2\end{array}$=T $\begin{array}{r}6\\+3\\\hline 8... \end{array}$=U $\begin{array}{r}3\\+3\\\hline 6\end{array}$=Y

9.

What kinds of keys won't open a door?

M O N K E Y S ,   T U R K E Y S ,
12 9 3 8 14 4 16   17 6 19 8 14 16

A N D   D O N K E Y S
18 3   13 9 3 8 14 16

$\begin{array}{r}9\\+9\\\hline 18\end{array}$=A $\begin{array}{r}6\\+7\\\hline 13\end{array}$=D $\begin{array}{r}4\\+4\\\hline 8\end{array}$=E $\begin{array}{r}3\\+0\\\hline 3\end{array}$=E $\begin{array}{r}2\\+6\\\hline 8\end{array}$=K $\begin{array}{r}5\\+7\\\hline 12\end{array}$=M $\begin{array}{r}8\\+0\\\hline 8\end{array}$=N $\begin{array}{r}0\\+3\\\hline 3\end{array}$=N

$\begin{array}{r}4\\+5\\\hline 9\end{array}$=O $\begin{array}{r}10\\+9\\\hline 19\end{array}$=O $\begin{array}{r}4\\+2\\\hline 6\end{array}$=S $\begin{array}{r}10\\+7\\\hline 17\end{array}$=T $\begin{array}{r}8\\+6\\\hline 14\end{array}$=Y

$\begin{array}{r}5\\+4\\\hline 9\end{array}$=O $\begin{array}{r}6\\+10\\\hline 16\end{array}$=R

13.

What colors would you paint the sun and wind?

T H E   S U N   R O S E   A N D
6 4 16   18 7 13   10 2 7 16   10 13 8

T H E   W I N D   B L U E
6 4 16   14 9 13 8   9 11 12 16

$\begin{array}{r}4\\+6\\\hline 10\end{array}$=A $\begin{array}{r}4\\+7\\\hline 11\end{array}$=B $\begin{array}{r}6\\+2\\\hline 8\end{array}$=D $\begin{array}{r}5\\+1\\\hline 6\end{array}$=E $\begin{array}{r}8\\+1\\\hline 9\end{array}$=E $\begin{array}{r}3\\+1\\\hline 4\end{array}$=H $\begin{array}{r}9\\+10\\\hline 19\end{array}$=I $\begin{array}{r}4\\+2\\\hline 6\end{array}$=T

$\begin{array}{r}7\\+6\\\hline 13\end{array}$=L $\begin{array}{r}2\\+0\\\hline 2\end{array}$=N $\begin{array}{r}10\\+8\\\hline 18\end{array}$=O $\begin{array}{r}7\\+0\\\hline 7\end{array}$=S

$\begin{array}{r}8\\+7\\\hline 15\end{array}$=U $\begin{array}{r}5\\+7\\\hline 12\end{array}$=U $\begin{array}{r}8\\+6\\\hline 14\end{array}$=W

17. How do locomotives hear?

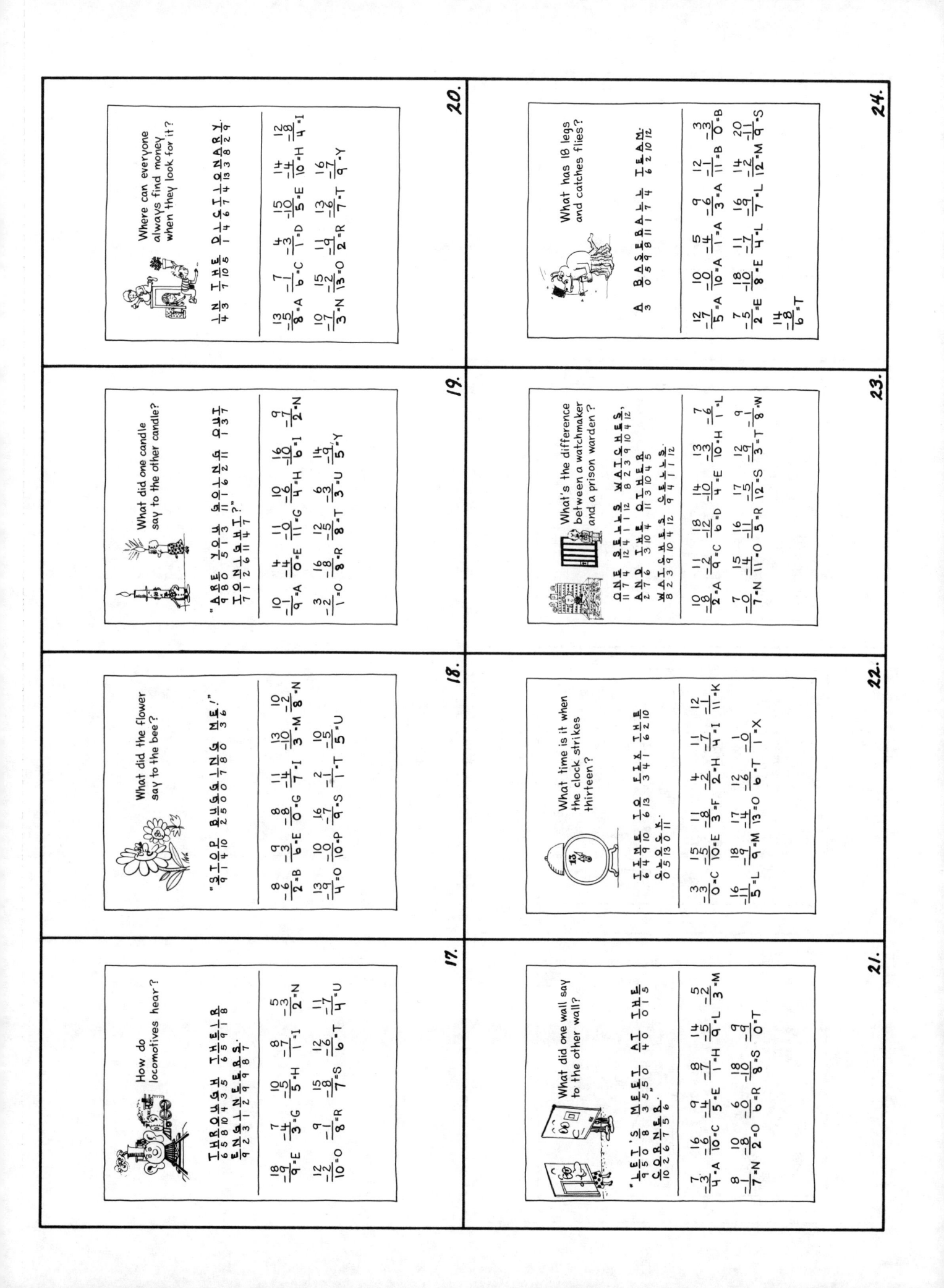

THROUGH THEIR
6 5 8 10 4 3 5 6 5 9 1 8
ENGINEERS.
9 2 3 1 2 9 9 8 7

$\frac{18}{-9}$=E $\frac{7}{-4}$=G $\frac{10}{-5}$=H $\frac{8}{-7}$=I $\frac{5}{-3}$=N $\frac{7}{-5}$=U

$\frac{12}{-2}$=O $\frac{9}{-1}$=R $\frac{15}{-7}$=S $\frac{12}{-6}$=T $\frac{9}{-4}$=U

18. What did the flower say to the bee?

"STOP BUGGING ME!"
9 1 4 10 2 5 0 0 7 8 0 3 6

$\frac{8}{-6}$=B $\frac{9}{-3}$=E $\frac{11}{-8}$=G $\frac{7}{-1}$=I $\frac{13}{-10}$=M $\frac{10}{-2}$=N

$\frac{13}{-4}$=O $\frac{10}{-10}$=P $\frac{16}{-7}$=S $\frac{2}{-1}$=T $\frac{10}{-5}$=U

19. What did one candle say to the other candle?

"ARE YOU GOING OUT
4 8 0 5 1 3 11 6 2 11 1 3 7
TONIGHT?"
7 1 2 6 11 4 7

$\frac{10}{-1}$=A $\frac{4}{-0}$=E $\frac{11}{-0}$=G $\frac{10}{-6}$=I $\frac{16}{-10}$=L $\frac{9}{-7}$=N

$\frac{3}{-2}$=O $\frac{18}{-8}$=R $\frac{12}{-5}$=T $\frac{9}{-3}$=U $\frac{14}{-9}$=Y

20. Where can everyone always find money when they look for it?

IN THE DICTIONARY.
4 3 7 10 5 1 4 6 7 4 13 3 8 2 9

$\frac{13}{-5}$=A $\frac{7}{-1}$=C $\frac{4}{-3}$=D $\frac{15}{-10}$=E $\frac{14}{-4}$=H $\frac{12}{-8}$=I

$\frac{10}{-3}$=N $\frac{15}{-13}$=O $\frac{11}{-9}$=R $\frac{13}{-6}$=T $\frac{16}{-7}$=Y

21. What did one wall say to the other wall?

"LET'S MEET AT THE
9 5 5 0 4 0 0 5
CORNER."
10 2 6 7 5 6

$\frac{7}{-3}$=A $\frac{16}{-6}$=C $\frac{9}{-4}$=E $\frac{8}{-7}$=H $\frac{14}{-5}$=L $\frac{5}{-2}$=M

$\frac{8}{-7}$=N $\frac{10}{-8}$=O $\frac{6}{-0}$=R $\frac{18}{-9}$=S $\frac{0}{-0}$=T

22. What time is it when the clock strikes thirteen?

TIME TO FIX THE
6 4 9 10 6 13 3 4 1 6 2 10
CLOCK.
0 5 13 0 11

$\frac{3}{-3}$=C $\frac{15}{-5}$=E $\frac{11}{-8}$=F $\frac{4}{-2}$=H $\frac{7}{-7}$=I $\frac{12}{-1}$=K

$\frac{16}{-5}$=L $\frac{18}{-9}$=M $\frac{12}{-6}$=O $\frac{9}{-0}$=T $\frac{2}{-1}$=X

23. What's the difference between a watchmaker and a prison warden?

ONE SELLS WATCHES,
11 7 4 12 4 11 2 8 2 3 9 10 4 12
AND THE OTHER
2 7 6 3 10 4 11 3 10 4 5
WATCHES CELLS.
8 2 3 10 4 12 9 4 11 12

$\frac{10}{-8}$=A $\frac{11}{-2}$=C $\frac{18}{-4}$=D $\frac{14}{-10}$=E $\frac{13}{-6}$=H $\frac{7}{-1}$=L

$\frac{7}{-0}$=N $\frac{15}{-11}$=O $\frac{16}{-5}$=R $\frac{17}{-12}$=S $\frac{12}{-3}$=T $\frac{9}{-8}$=W

24. What has 18 legs and catches flies?

A BASEBALL TEAM.
3 0 5 9 8 11 7 4 6 2 10 12

$\frac{12}{-5}$=A $\frac{10}{-10}$=A $\frac{5}{-4}$=A $\frac{9}{-6}$=A $\frac{12}{-3}$=B

$\frac{7}{-2}$=E $\frac{18}{-8}$=E $\frac{11}{-7}$=L $\frac{16}{-2}$=M $\frac{20}{-11}$=S

$\frac{14}{-8}$=T

25. Why do dragons sleep in the daytime?

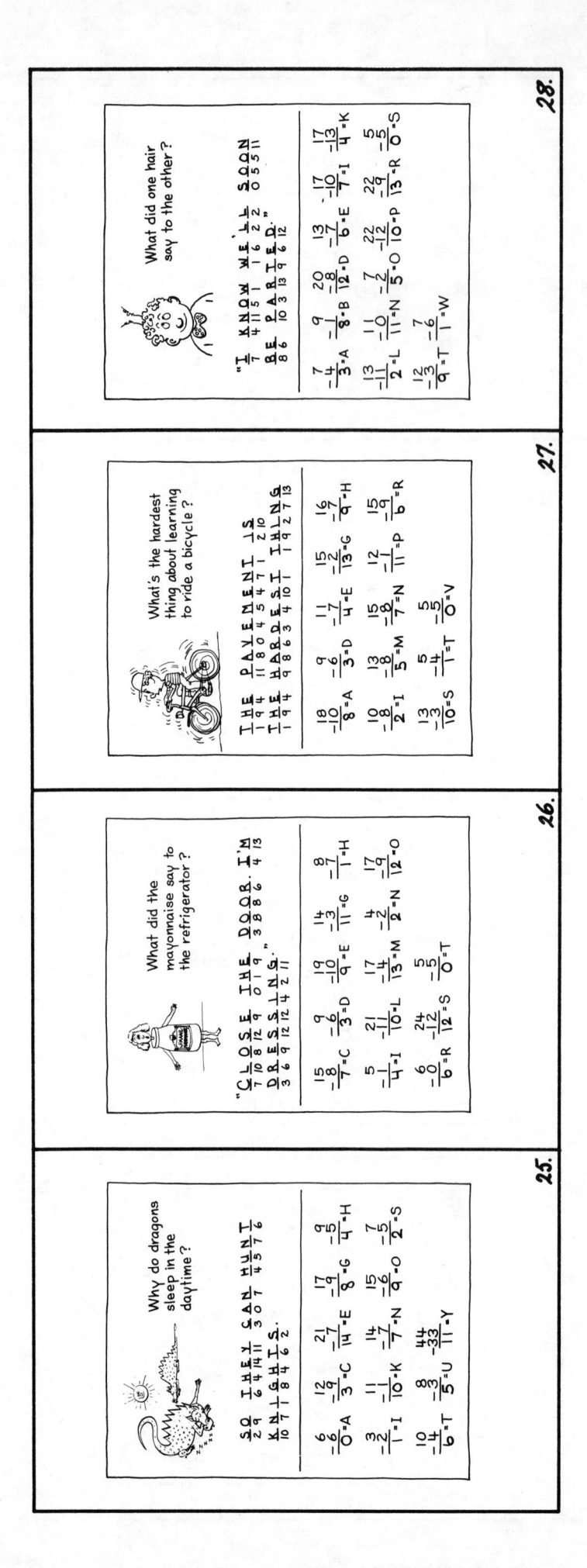

"S O   T H E Y   C A N   H U N T
 2 9   6 4 14 11   3 0 7   4 5 7 6

K N I G H T S ."
10 7 1 8 4 6 2

$\dfrac{6}{-6}$=A $\dfrac{12}{-9}$=C $\dfrac{21}{-7}$=E $\dfrac{17}{-9}$=G $\dfrac{9}{-5}$=H

$\dfrac{3}{-2}$=I $\dfrac{11}{-1}$=K $\dfrac{14}{-7}$=N $\dfrac{15}{-9}$=O $\dfrac{7}{-5}$=S

$\dfrac{10}{-4}$=T $\dfrac{8}{-3}$=U $\dfrac{44}{-33}$=Y

26. What did the mayonnaise say to the refrigerator?

"C L O S E   T H E   D O O R .   I'M
 7 10 8 12 9   0 1 9   3 8 8 6   4 13

D R E S S I N G ."
3 6 9 12 12 4 2 11

$\dfrac{15}{-8}$=C $\dfrac{9}{-6}$=D $\dfrac{19}{-10}$=E $\dfrac{14}{-3}$=G $\dfrac{8}{-7}$=H

$\dfrac{5}{-1}$=I $\dfrac{21}{-8}$=L $\dfrac{17}{-4}$=L $\dfrac{14}{-2}$=N $\dfrac{17}{-9}$=O

$\dfrac{6}{-0}$=R $\dfrac{24}{-12}$=S $\dfrac{5}{-5}$=T

27. What's the hardest thing about learning to ride a bicycle?

T H E   P A V E M E N T   I S
1 4   1 1 8 0 4 5 4 7 1   2 10

T H E   H A R D E S T   T H I N G
9 8 6 3 4 10 1   9 2 7 13

$\dfrac{18}{-10}$=A $\dfrac{9}{-6}$=D $\dfrac{11}{-7}$=E $\dfrac{15}{-2}$=G $\dfrac{16}{-7}$=H

$\dfrac{10}{-8}$=I $\dfrac{13}{-8}$=M $\dfrac{15}{-7}$=N $\dfrac{12}{-1}$=P $\dfrac{15}{-9}$=R

$\dfrac{13}{-3}$=S $\dfrac{5}{-4}$=T $\dfrac{5}{-5}$=V

28. What did one hair say to the other?

"I   K N O W   W E' L L   S O O N
 7   4 11 5 1   1 6 2 2   0 5 5 11

B E   P A R T E D ."
8 6   10 3 13 9 6 12

$\dfrac{7}{-4}$=A $\dfrac{9}{-8}$=B $\dfrac{20}{-8}$=D $\dfrac{13}{-7}$=E $\dfrac{17}{-7}$=I $\dfrac{17}{-13}$=K

$\dfrac{13}{-11}$=L $\dfrac{10}{-2}$=N $\dfrac{7}{-2}$=O $\dfrac{22}{-12}$=P $\dfrac{22}{-9}$=R $\dfrac{5}{-0}$=S

$\dfrac{12}{-3}$=T $\dfrac{7}{-6}$=W